CAN YOU SOLVE THESE?

Mathematical problems to
test your thinking powers.
SERIES No. 1

DAVID WELLS

TARQUIN PUBLICATIONS

Most of the problems in this book first
appeared, sometimes in a different form,
in issues 1, 2 and 3 of 'The Problem Solver',
written and edited by David Wells.

An up-to-date catalogue can be obtained from the
publisher at the address below.

© 1995:	David Wells	Tarquin Publications
© 1982:	Previous Edition	Stradbroke
I.S.B.N.:	0 906212 22 7	Diss
Design:	Wilson Smith	Norfolk IP21 5JP
Printing:	Ancient House Press, Ipswich	England

PROBLEM SOLVING

Archimedes was enjoying his bath when he suddenly hit
upon the solution to the problem of Hiero's Golden Crown.
Flushed with delight, not embarrassment, he ran naked into
the street shouting, 'Eureka, Eureka! I have found it, I have
found it!' We do not recommend that you follow
Archimedes' example to the letter, but we do hope that you
get a kick out of many of these problems.

They are all based on mathematical ideas and
mathematical thinking, but don't worry if you have forgotten
most of the mathematics you ever knew. Common-sense
and ingenuity, perseverance and a flash of insight will be
much more useful than book-knowledge.

Some of these problems are easy, some are difficult, but
they have all been chosen for your enjoyment, to stimulate
your interest in mathematical ideas, and as a starting point
for problems and investigations of your own.

If you get really stuck, there are clues to most of the
problems in the HINTS section, and answers to most of
them in SOLUTIONS, but not to all of them. Just as real
mathematicians always have new problems to tackle and
loose ends to tie up, so we leave a few of these problems
entirely up to you!

Good problem solving!

David Wells

These two triangles have their sides as given. Which has the larger area?

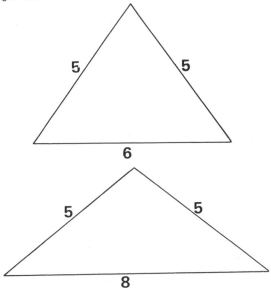

Can you cross out twelve points arranged like this by drawing 5 straight lines, *without* taking your pen off the paper *and* ending up where you started?

BLOCKS OF FOUR

1	6	3	
2	1	0	

These SIX numbers have been arranged so that each complete block of four adds up to exactly 10.

4	2	5	1
3			
3			
1			

Can you complete this square so that each block of four adds up to exactly 10?

4		4	
1		2	3
		5	0

Can this square be completed using the same rule?

Can you rearrange the seven numbers above, so that the square can be completed?

What if you have more or less than 7 numbers to start with?

CROSS-NUMBERS

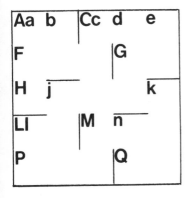

No answer starts with a nought

ACROSS
A: a perfect square
C: a perfect square
F: a perfect square
G: a perfect cube
H: C across, squared
L: twice A across
M: 1000 — j
P: Divisible exactly by 5, 7 and 8
Q: $\frac{1}{2}$M—2

DOWN
a: a prime number
b: a perfect square and a perfect cube
c: a perfect square + 10
d: a perfect square
e: a perfect square
j: 1000 — M
k: 8n
l: has 6 different factors including itself and 1
n: one eighth of k

3

7 does not divide exactly into 9, or into 99, or into 999. What numbers made up entirely of 9s, can be divided by 7 with no remainder? What is the result of each division?

Which numbers composed of 9s can be divided exactly by 13? By 17? By 23? By 24?

Which numbers made up from 4s can be divided exactly by 7?

What are the next three numbers, from top to bottom, in this treble sequence?

What happens to the sequence if it is continued very much farther?

3	11	-5	27		
8	1	15	-13		
6	5	7	3		

The diameter of the smaller circle is 5 cms and of the larger 7 cms.

Which is the larger, the area of the inside circle, or the area between them?

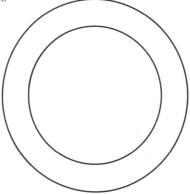

Which is larger, the vertically shaded area or the horizontally shaded area?

The diameters of the circles are 6 cm, 4 cm, 4 cm and 2 cm.

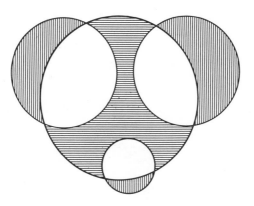

The sum of two numbers is 12, and their product is 27. What are the two numbers?

The sum of two numbers is 12, and their product is 22.31. What are the two numbers?

This time they add up to 12, but their product is 22. Find the two numbers as accurately as you can.

In how many different ways can the word THINKER be read from left to right in this diagram, each letter connecting to its neighbour along one of the marked lines.

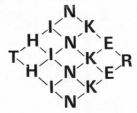

As you take a photograph of a train hurtling from London to Bristol at 120 mph, some bits of the train are actually moving in the opposite direction, back to London. Which bits?

This diagram shows how a regular ten-sided figure can be divided into ten diamonds (also called rhombuses or rhombi!)

How many diamonds can a regular 14 sided figure be divided into using the same system?

How many diamonds will there be in a regular 6 sided, 18 sided and 20 sided figure?

Are there any number of sides for which it cannot be done?

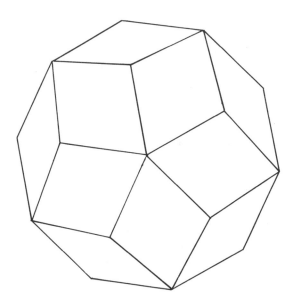

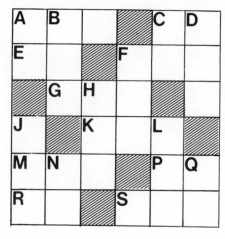

CLUES ACROSS

A	205 + 214
C	3 x 21
E	½ of 46
F	400 — 153
G	5 x 110
K	982 — 73
M	4 x 5 x 6
P	7 x 7
R	275 ÷ 5
S	500 + 160 — 9

CLUES DOWN

A	6 x 7
B	79 + 56
C	4 x 4 x 4
D	400 — 28
F	1000 ÷ 5
H	632 — 42
J	46 + 22 + 47
L	315 x 3
N	100 ÷ 4
Q	13 x 7

Each side of the hexagon on the left is two units long. It can be divided into four equal smaller hexagons in only 6 pieces, as the right-hand diagram shows.

If you start with a larger hexagon whose sides are three units, how many smaller hexagons will you get, in how many pieces?

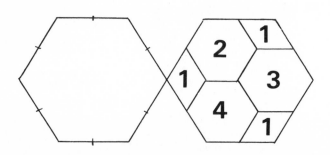

This is a regular hexagon divided into six identical parts.

Copy this figure, cut out the six pieces and use them to make two regular pentagons. (A regular pentagon has five equal sides and five equal angles.)

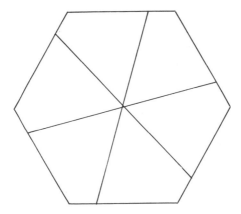

You have 2 parents, 4 grandparents, 8 great grandparents, 16 great grandparents . . . and so on. If each generation on average lives 25 years further back in time, how many great great great great . . . great grandparents did you have a thousand years ago? Compare your answer to the total number of people in the world at that time.

How can you pour liquid into a cylindrical can until it is half full, if there are no marks on the inside or outside of the can?

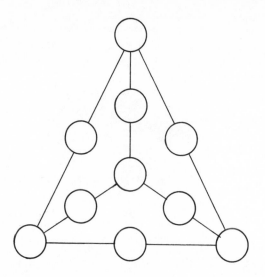

Here are six rows of three circles each.

Place the numbers zero to nine in the circles, so that the rows add up to 8, 9, 10, 11, 12 and 13.

They do not have to add to these totals in any particular order.

Which numbers less than 100 have exactly three factors, including themselves and one?

What number less than 100 has the largest number of factors? (ALL factors count, not just prime factors.)

21

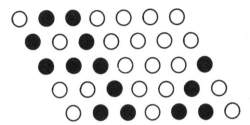

This pattern starts from the top row and works downwards. Each row is worked out from the line above, by using the same rule.

What is that rule?

What happens if you add more rows to the bottom, how will they be coloured?

Will the pattern ever repeat?

If you start with a line of circles and then work down like this in a triangular pattern, can you predict what colour the bottom circle will be simply by looking at the top row?

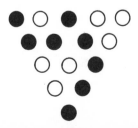

HOW FAST CAN YOU COUNT?

It is very easy to count to ten in just a few seconds. If you count quickly you might be able to reach 40 or even 50 in ten seconds.

The problem is to estimate how long it would take to count to a million.

Don't forget that it takes longer to say 609 than it does 69, and much longer to say 609,746. However, with some thought and trials with a watch or a stop-watch, you should be able to make a reasonable estimate!

This diagram shows how twenty prisoners are arranged in eight cells, with six prisoners in each row of three cells. Rearrange the prisoners so that there are seven in each row of three.

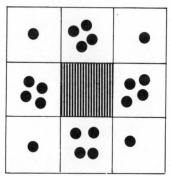

I think of three numbers; (Call them P, Q and R) I work out that:—

$$2P + Q + R = 22$$
$$P + 2Q + R = 20$$
$$P + Q + 2R = 18$$

What numbers did I think of?

BOOTSTRAPS CROSS-NUMBER

Pulling yourself up by your own bootstraps is proverbially difficult. This cross-number is tricky, because many clues depend on the answers to others — but never fear, it can be done!

No answer starts with a zero

CLUES ACROSS
A $J - 1$
C $2f$
F f^2
H $Q + R$
J $A + 1$
M $\frac{1}{4}Q$
N $g + j$
Q $k + p$
R $\frac{1}{4}J$

CLUES DOWN
b $A - \sqrt{R}$
d $5b$
e $Q - R$
f $F \div f$
g $3j$
j $\frac{1}{3}$ of g
k $Q - p$
m $g \div 27$
p $Q - k$

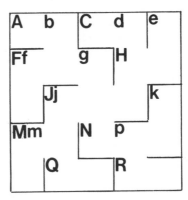

26

CODEBREAKING

The words of a song are printed here, but you will have to crack the code first to be able to read it!

T FPU YTFY BYPR T CPP ZDH FD VZ,

VHUUPLXNZ, BYZ WJ T CD CYZ, BYPR

T'J VPCTOP ZDH? TU'C DRNZ NDKP

WRO UYWU TC WNN, UYWU JWQPC

JP WEU UYP BWZ T OD, TU'C DRNZ

NDKP WRO UYWU TC WNN, VHU TU'C

CD YWLO NDKTRF ZDH.

This diagram shows nine lines and nine points with exactly three points on every line and three lines through every point.

Find another arrangement of nine lines and nine points with the same property.

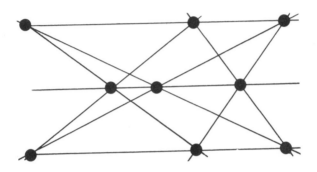

How many triangles are there in each of these figures? What figure comes next in the sequence and how many triangles will there be in it?

Find the smallest rectangle which can be made up from pieces this shape, only,

without ever putting two of the pieces together like this to made a 3 x 2 rectangle.

What is 777777 x 999999?

(Warning! You are not supposed to work it out by one very long sum; look for a pattern instead.)

What is special about 2 x 3529411764705882?

Find out everything else you can which is peculiar about the 16-digit number.

32 When a plane crashed it was equi-distant from Montebourg and Carentan.

It was also twice as far from the smaller of the Isles St. Marcouf as from Utah beach.

Where did it crash?

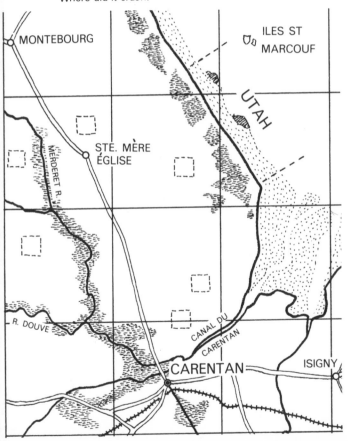

What is the smallest rectangle which can be made by using jigsaw pieces of these two shapes only?

What is the next smallest such rectangle?

Is it possible to make all sufficiently large rectangles by using these shapes only?

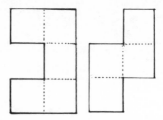

Place arithmetical signs between these numbers so that they make a correct sum:

$$9\ 8\ 7\ 6\ 5\ 4\ 3\ 2 = 1$$

You can use +, —, × and ÷ and also brackets if necessary.

Three numbers multipled together come to 479491.
Find the three numbers.

At a special meeting, everyone shakes hands exactly once with each other person present. Altogether there are 45 handshakes.
How many people attended the meeting?

 Copy these five figures by tracing them, and then fit them together to make a cross.

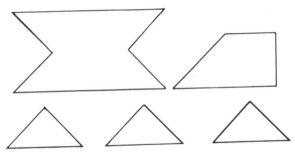

 If this addition sum went on for ever, what do you think the answer might be?

$\frac{1}{3} + \frac{1}{9} + \frac{1}{27} + \frac{1}{81} + \frac{1}{243} + \frac{1}{729} + \ldots$

SHEEP MAY SAFELY GRAZE, AND HAVE PLENTY TO EAT!

A farmer has 100 metres of wire fencing which he wants to use to surround as much area as possible against a long straight wall, which is shaded in the diagram below.

The figure shows three rather poor attempts to solve the problem.
Can you find better solutions? Can you find the best possible solution?

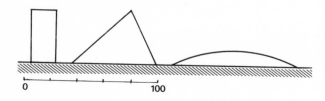

TOP FORTY

In the BBC Top Forty charts which are compiled each week, usually a few records rise up the chart, compared to last week, a few drop down and some stay put. Some records leave the chart and some new ones join it.

What is the largest number of records which were in the chart last week, which could have risen up the chart by this week?

41 If you turn a left-handed glove inside out, will it then be left-handed or right-handed?

42 Two ladders are propped up vertically in a narrow passageway between two vertical buildings. The ends of the ladders are 8 metres and 4 metres above the pavement.

Find the height of the point where they cross, T, above the pavement.

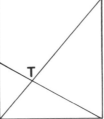

43 What is the smallest possible number of children in the Jones family, if each child has at least one brother and at least one sister?

Prove, without solving the equation, that if one solution of,

$$X^2 - Bx = A$$

is x = s, then another solution is x = B — s

 EQUAL AREAS

Construct six points A, B, C, D, E, F such that the three quadrilaterals
     ABDE
     ACDF
     BCEF

are all equal in area, but not the same shape.

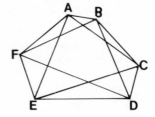

 Rearrange all these words to make a true sentence, adding punctuation where necessary:

every or parallelogram rectangle true either is that not it a a kite quadrilateral a is

Here are the 8 volumes of Anne's Encyclopaedia on her bookshelf. However, they need rearranging in order.

If taking a book from the shelf, pushing some books to one side and putting the book back counts as one move, what is the smallest number of moves needed to arrange the set correctly, from 1 to 8, left to right?

=	=	=	=	=	=	=	=
8	4	2	5	1	7	3	6

This was originally a magic square: (each of the rows and columns and the two diagonals added up to the same total.)

However, it has been spoilt by swopping three pairs of numbers.

Discover which pairs of numbers have been exchanged, and restore the original square.

16	2	3	13
6	11	10	7
9	8	5	12
1	14	15	4

Which of these numbers is the larger: 2^{30} or 3^{20}?

50 What is the smallest rectangular area of wrapping paper which can be used to wrap this parcel?

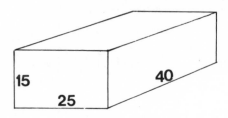

51 23 cubes have been fitted together exactly face to face, to nearly make one complete loop.

What is the smallest number of extra cubes needed to complete it?

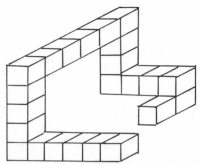

A QUESTION OF SCALE

This figure shows two maps of the same area of country, but one is double the scale of the other.

One map lies on top of the other, and there is a point on the top map which is exactly above its corresponding point on the bottom map.

Can you find it?

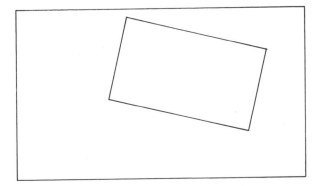

A bank clerk counts a very large amount of small change at the end of each day's business. If he works 250 days a year, how many times in a year would he expect the total to be an exact number of pounds?

A RANDOM WALK

The Diceman moves by throwing a dice at every intersection. After a 1 or 2 he turns right; after 3 or 4 he turns left, and after 5 or 6 he crosses straight over.

The diagram shows the results of throwing 1, 4, 4, 6, 2, 6, 1, 2, 3.

Either by throwing a dice many times or by using your computer to do this for you, investigate long journeys by the Diceman.

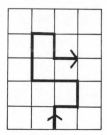

If two of these numbers are crossed out, the sums of the remaining numbers across all the rows and down all the columns will be multiples of 5.

Which two should be crossed out?

1	2	4	8
5	3	2	3
7	7	1	6
2	6	3	9

Smith, Jones, Brown, John, Peter and David are three teachers at Gruntly School, teaching different subjects. Smith teaches French, Brown gives the science teacher a lift home in his car, John teaches art and David cannot stand the smell of the chemistry laboratories.

Match the first and second names of each teacher.

This rectangle has been folded, from corner to corner so that X is on top of Y.

If the total area of the two wings marked W is equal to area C, what shape is the rectangle?

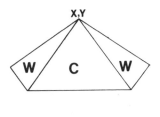

If you double 6, then double it again . . . and again . . . and again . . . like this:

　　　6　　12　　24　　48　　96　　192 . . .

you will never get to a number which is a perfect square. Explain why.

59

Here are three examples of ways of subtracting 876 from 1225. The method on the left is very common.

The other two methods are quite unusual. Your problem is to discover *how* they work, and *why* they work.

$$
\begin{array}{r}
1225 \\
876 \\
\hline
349
\end{array}
\qquad
\begin{array}{r}
1225 \\
876 \\
\hline
226 \\
123 \\
\hline
349
\end{array}
\qquad
\begin{array}{r}
1225 \\
8775 \\
\hline
876 \\
9651 \\
\hline
349
\end{array}
$$

60

MISSING LINKS

Each of these mathematical sentences was true until some of the symbols were taken out.

The missing symbols are in the right-hand margin, NOT in any particular order. Where should they be replaced?

24	13	21	=	16	— +
40	8	3	=	8	+ ÷
6	4	6	=	30	× +
64	8	2	=	16	() ÷ ÷
2	3	4 + 1	=	25	() () + ×

61

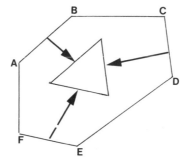

This hexagon A, B, C, D, E, F is rather special, because the sides AB, CD, EF will form a triangle by themselves, simply by being moved *without rotation* as the diagram shows.

Explain why BC, DE and FA must also form a triangle in the same way — by being slid across the page without rotation.

PERFECT SQUARES

62

$(p-10)(p+14)$

Find all the positive integers which make this expression a perfect square.

(To set you going, 0 counts as a perfect square for this problem!)

$(p-6)(p+14)$

Similar problem to investigate!

63

Is it possible to find three numbers, a, b, c, none of which is zero or a perfect square, such that:

$$\sqrt{a} + \sqrt{b} = \sqrt{c}?$$

64

Many numbers can be written as the difference of two perfect squares. For example:

$$5 = 9 - 4 \qquad 20 = 36 - 16 \qquad 21 = 25 - 4$$

Which numbers *cannot* be written as the difference of two perfect squares?

The following numbers have to be fitted into this crossnumber diagram.

2815, 4672, 7135, 9612, 172, 237, 279, 431, 517, 714, 16, 37, 71, 81.

Be careful, because there are several ways to fit some of them in, but only one way to fit them all in.

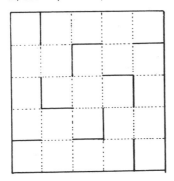

Ron Gasper, the long distance runner, has a great idea for increasing his speed. He used to run in shoes with ordinary soles, as the diagram on the left shows, but now he has changed to shoes with an extra 2 cm of rubber on each sole, as you can see on the right.

As that diagram shows, each of his strides is now about 1 cm longer than before.

Well — is Ron Gasper's idea really so great?

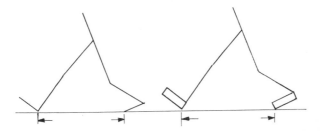

I am using a rubber band to hold some pencils together. To make sure that the band is tight, it is doubled round the pencils. If I now straighten out the band as much as I can, how many twists will finally be left in it?

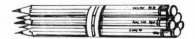

Here are two triangles, with their vertices on the intersections of a square grid. The first has two points inside and three points on its top edge, (not counting its vertices.)

The second has only one point inside and one point on an edge.

How many different *shaped* triangles can you draw with their vertices on the intersections of a square grid, which have only one point inside them and *no* points on their edges?

You can imagine that the grid is as large as you like, though of course we can only draw a small bit of it!

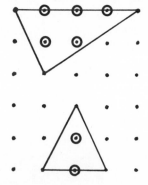

69 DISSECTIONS

This is a regular hexagon.

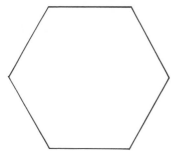

How can it be cut into two pieces which will make a parallelogram?

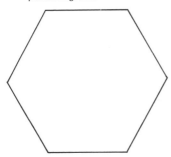

How can it be cut into three pieces which will make a rhombus?

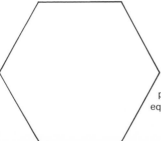

How can it be cut into four pieces which will make two equilateral triangles?

70 Eight years ago my father was three times as old as I shall be in five years time. When I was born he was 41 years old. How old am I now?

71 The first few of these sums are easy to work out, but as they get longer, they get much harder. Find a rule for working them out *without* having to add them up one number at a time!

1 + 3	= 4
1 + 3 + 9	= 13
1 + 3 + 9 + 27	= 40
1 + 3 + 9 + 27 + 81	= ?
1 + 3 + 9 + 27 + 81 + 243	= ?
....	....

72 What might this sum equal if it 'carried on for ever'?
$1 - \frac{1}{2} + \frac{1}{4} - \frac{1}{8} + \frac{1}{16} - \frac{1}{32} + \ldots$

Both of these quadrilaterals have sides of length 7, 5, 5, 1 in that order, but their areas are not the same.

What is the area of the largest possible quadrilateral with those sides in the same order?

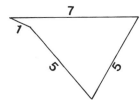

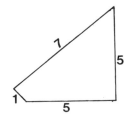

These five wheels all rotate in contact with each other, without slipping. The left-hand wheel is 40 cms diameter and rotates 12 times per minute. The right-hand wheel is 30 cms diameter. How fast does is rotate?

What is the diameter of the smallest wheel if it rotates once every second?

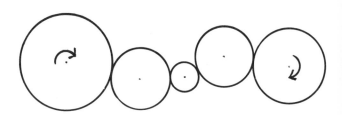

75 An examination question is marked out of five, and only whole numbers of marks are given. Five pupils answer the question, and the mode of their marks is one more than the median, which is one more than the average mark.

What marks did each of the five pupils get?

76 This diagram shows a quadrilateral, ABCD, with each vertex joined to a point marked X.

How can this diagram be used to draw other quadrilaterals which are the same shape as ABCD, but of different sizes?

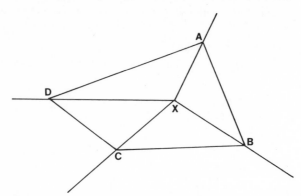

77 MULTIPLE CROSS

Each answer is a multiple of the number appearing in its first square. For example, the number along the top line is a multiple of 17.

17	17	9	35
24		96	
29	29	13	
	7		

78

What do you think should be the next three numbers in each of these sequences?

1	3	7	15	31	63	—	—	—	
0	1	4	10	20	35	—	—	—	
1	1	3	5	11	21	43	—	—	—
1	1	1	3	5	9	17	—	—	—

79

The diagrams each show a non equilateral triangle with an equilateral triangle drawn with each vertex on one side.

Is it always possible to draw such an equilateral triangle and how would you construct it?

How many different equilateral triangles can be drawn in any triangle?

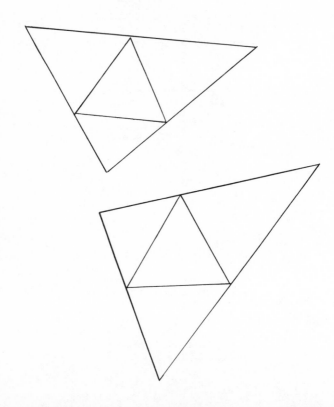

In each direction the numbered arrows show the totals of the rows in three directions. Can you work out what numbers the letters stand for?

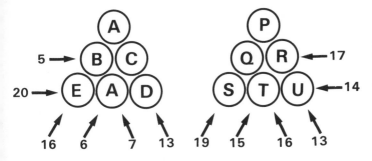

Find the numbers the letters MAGIC stand for in this magic star. Each row adds up to 28.

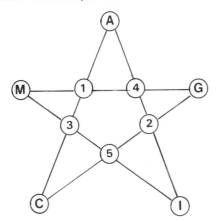

The diagram shows three different ways of making your first two moves.

In the game of REACH you can make any of three moves; you can *either* move 3 spaces to the right and 2 up; *or* you can move 4 spaces to the right and 5 up, *or* just 5 right.

Starting at the black dot, can you REACH a point which is 30 spaces to the right and 40 spaces up? If you can, what is the smallest number of moves you have to take?

If you cannot, what is the nearest point you can REACH?

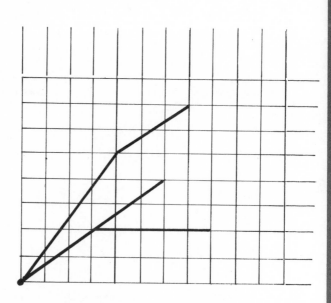

There is only one number, apart from 1, which divides exactly into each of these numbers. What is it?

163231   152057   135749

In these sums, some of the figures have been carelessly blotted out. Discover what they should be!

●2●
x ___7
1●7●

3●
- ●9
___9

●2●
8)●9●

This diagram shows a part of a large irregular tessellation in which three polygons only meet at each vertex.

What is the average number of sides per polygon in any very large tessellation of this kind?

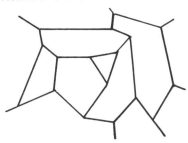

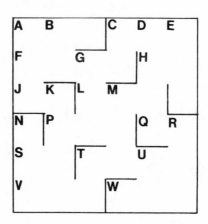

CLUES ACROSS
A 12 × 12
C 400 — 76
F 33 × 34
H 583 ÷ 11
J 8 + 4 + 2 + 1
L 2 × 2 × 2 × 3 × 7
P 7 × 7 × 7
Q 93 ÷ 3
S 3 × 3 × 3
T 1472 — 390
V 40 + 42 + 44 + 46
W 1 × 2 × 3 × 4 × 5 × 6

CLUES DOWN
A 3 × 37
B 369 ÷ 9
C 2 × 2 × 2 × 2 × 2
D 369 × 7
E 143 + 144 + 145
G 856 ÷ 4
K 4729 + 648
M 8820 ÷ 14
N 22 × 22 ÷ 4
R 1 × 2 × 3 × 4 × 5
T 8 × 9 ÷ 6
U 19 + 20 + 21 + 22

This diagram shows the postal districts, including their numbers, in the city of Lantwen. Unfortunately people get rather confused when they discover that postal district 10 is next to district 3 and district 7 is next to district 1. The Postmaster has decided therefore to renumber the districts, to make the differences between the numbers of adjacent districts as small as possible.

How would you advise the Postmaster?

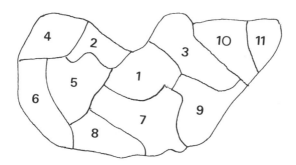

Find both solutions of the equation: $T^2 - 6T + 4 = 0$ by completing this argument:

"The sum of the roots is 6. Therefore the average of the two roots is 3, and the roots can be written in the form $3 + p$ and $3 - p$.

But the product of the roots is 4. Therefore . . ."

Mary and Peter were arguing about the shape they would get by slicing a circular cone with a flat knife, as in the diagram. 'It will be an oval,' said Peter. 'Of course it will,' said Mary, 'with the wider part at the top because there the knife will cut the edge nearly at right angles.'

'I disagree,' replied Peter, 'it will be wider at the bottom, because that is where the cone is widest.' Who was right, Mary or Peter?

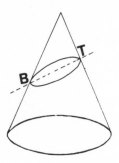

Which is greater: $1 + \sqrt{34410}$ or $\sqrt{34782}$?

I think of three numbers. Call them F, G and H. I discover, that

$$F + G + H = 10, \qquad F \times G \times H = 30$$
$$F \times G + G \times H + H \times F = 31$$

Find the numbers I thought of.

Here is one way to share five bars of chocolate equally between 8 people:

The next diagram shows one way to share 3 equally between 5 people.

Using the same method, how could 6 bars be shared equally between 8 people?

 Share these numbers into two sets, so that the difference between the sums of the numbers in each set is as small as possible:

17. 4. 67. 5. 24. 31. 46. 19.

As in problem 12 this figure shows how a regular 10-sided figure, (a regular decagon,) can be dissected into 10 rhombuses.

Starting with this idea, dissect a regular decagon into two identical regular pentagons plus one, smaller, regular decagon.

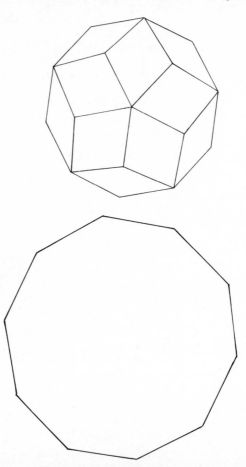

A gardener wishes to plant as many rose bushes as he can in a rectangular rose bed, which is 20 metres long and 10 metres wide. Two rose bushes must not be less than 1 metre apart. Bushes must not be planted on the edge of the bed but they can be planted just inside the edge.

What is the largest number of rose bushes which the gardener can plant?

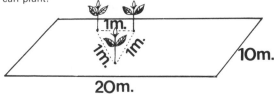

Mr. Dickinson is especially critical if any of his pupils gets the right answer to a problem despite making a mistake on the way. He is likely to exclaim, 'Anyone who gets the correct answer after one mistake, must have made at least two mistakes!!'

When is Mr. Dickinson's statement not true?

Have you ever seen anyone running along the pavement and placing their feet on the ground in this order: right foot, right foot, left foot, left foot, right, right, left, left, right . . .??

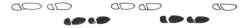

In this diagram, each letter stands for the size of the angle it marks. Explain why: A + B = P + Q + R + S

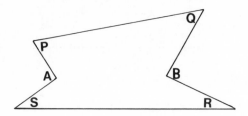

In this table each Y number, (except the first,) is one half of the previous X number; and each X number (except the first,) is 4 more than the difference between the previous X and Y numbers.

Is that clear?! Right — what happens to the numbers in each column when the table is continued for many more steps?

What happens if you start with other pairs of numbers?

X	Y
10	7
7	5
6	3.5
6.5	3
7.5	3.25
. . .	. . .
. . .	. . .
. . .	. . .
. . .	. . .
. . .	. . .

100

A cube is made out of wire, and three flies are crawling along the wires. The distances between any two flies is the same and it does not change as all three flies walk.

Note that the distance between the flies is measured *along* the edges of the cube and *not* across the faces.

Describe the movement of the flies and state the distance between them.

Is it possible for a fourth fly to join them under the same conditions?

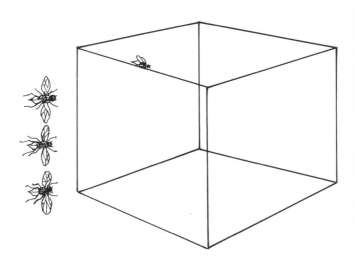

101

A knight moves across a square grid just as it would in a game of chess, except that it moves on the intersections and not inside the squares.

If it starts at (0, 0) what is the smallest number of moves it needs to reach the point (27, 5)?

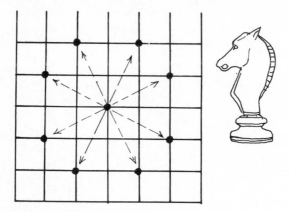

102

Every member of the Wisebuy Book Club is given a membership number. This is actually two letters of the alphabet followed by a two-digit number. How many members could the Book Club have before they run out of membership numbers?

UR12

This quadrilateral has been divided into four triangles, by choosing a point C and joining it to each of the four vertices.

However, in this example the four triangles are not equal in area to each other.

Is it possible to make them equal by choosing C somewhere else? If it is only possible with a quadrilateral of a different shape, what can you say about the necessary shape?

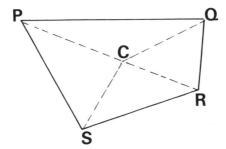

The sum of two numbers is 12 and their product is 30.

Without finding them first
Calculate the sum of their squares
Calculate the sum of their reciprocals.

The first image

Mr. Jones is five times as old as his son. How much older will he be when his son is twice his present age?

106

If the side AB of this rectangle is folded exactly onto the opposite side CD, one straight crease will be made. Suppose, however, that AB is folded onto CD, but shifted sideways, as shown by the dotted lines. How many creases will there now be? Is it possible to make this fold with a number of straight creases, or must the paper be distorted?

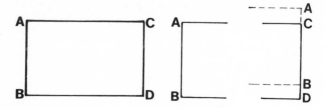

107

A cube is resting on an inclined plane so that it very nearly tips over, but not quite. It is rough enough not to slip.

How steep *must* the slope be if the cube is on the point of tipping, and how steep *could* the slope be if the cube is placed in a different orientation?

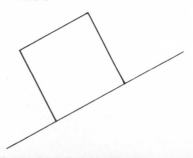

These lines form a spiral, which moves round and inwards towards a *limit point*.

Each vector is one half of the length of the previous vector, and at an angle of 60° to it. If the first vector is one unit long, find the position of the limit point.

How could the position of the limit point be calculated if each vector was 0.7 of the previous vector in length, and at an angle of 37° to it?

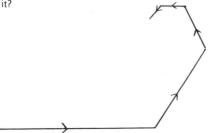

What does this sum equal?

$1 - 2 + 3 - 4 + 5 - 6 + 7 - 8 + 9 - 10 + 11 \ldots$

$\ldots - 98 + 99 - 100$

110

I have a large number of tiles, all the same shape and size. They are pentagons with angles of 100°, 130°, 130°, 40° and 140° in that order.
Can I use them to tessellate a floor?

140
40
100
130
130

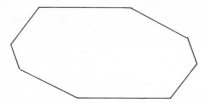

This is a rather special octagon. Its eight sides are parallel in opposite pairs.

In how many different ways can it be divided into six parallelograms?

A large reservoir is being emptied by 3 sets of pipes, all at the same level.

P — one pipe of diameter 60 cm
Q — two pipes of diameter 30 cm
R — three pipes of diameter 20 cm

Assuming that friction in the pipes is not important, which set of pipes is contributing most to emptying the reservoir?

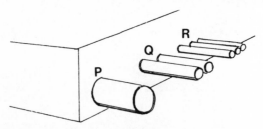

113

No answer starts with a zero

CLUES ACROSS		CLUES DOWN	
B	$b - 2$	a	C
C	a	b	$c\frac{1}{2}$
D	$\frac{1}{2}BK$	c	b^2
F	$e^2 \div J$	e	$3F$
J	$3e$	g	$2F$
K	$\sqrt{6L}$	h	$K^4 \div L^2$
L	$K^2 \div 6$	j	$16b$
P	$2K - C$	m	$J \div \sqrt{J}$
Q	$n \div 3$	n	$3Q$

a	Bb		Cc	
D		e	F	g
h	Jj			
K		L	m	n
P		Q		

114

If 6 matches are used to make a triangular pyramid, each end of each match touches two other matches. Arrange 16 matches, so that each end of each match meets three other matches.

115 MAGIC SQUARES

For your first trick, place the numbers 1 to 9 into this square so that each row and column and both the long diagonals add up to 15.

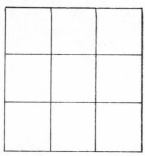

For your second trick place the numbers 1 to 25 into this larger square so that the whole square is magic, (every row and column and both the diagonals will add up to 65) *and* so that the 3 by 3 square in the centre is magic by itself. Difficult!

116

This triangle has been divided into four quadrilaterals and another triangle.

Is it possible to divide it into a number of quadrilaterals only, without marking any new vertices on the sides of the original triangle?

117

The operation P turns x into x +2; the operation Q turns x into $3x^{-1}$. Choose any number you like to start with and apply the operations P and Q alternately, many times. What answers do you get?

Investigate the results of starting with different numbers, or using slightly different operations, such as $x \rightarrow x+3$ and $x \rightarrow 10x^{-1}$.

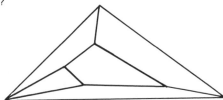

$$10 \xrightarrow{P} 12 \xrightarrow{Q} \frac{3}{12} \xrightarrow{P} \frac{27}{12} \xrightarrow{Q} \frac{36}{27} \rightarrow ????$$

Change every letter, don't go wrong,
the words, decoded are a song!

FZ FD VRT SHRIGQ DEVD F

MGGB ZH LBOG, SEGR F NRHS

MHQ KGQDVFR DEVD F'P

BHZFRA THO. SG OZGI DH

EVWG AHHI DFPGZ DHAGDEGQ,

LOD RHS F ZGG DEGP ZBFJ

VSVT, FD LQGVNZ PT EGVQD,

DH ZGG OZ JVQD, ZH ZVI

DH SVDKE AHHI BHWG

AH LVI.

119

You will need a large number of cardboard squares for this problem. How is it possible to construct a shape, a 'polyhedron', which has five squares meeting at every vertex?

(The word 'polyhedron' is in quotation marks because the answer is a rather weird shape, unlike ordinary polyhedrons.)

Another, similar, problem, with an even more curious solution: construct a 'polyhedron' which has six square faces meeting at each vertex.

120

I think of three numbers. Call them P, Q and R. P is the average of Q and R. R is twice the sum of P and Q. Explain why one of the numbers I thought of must be zero, and say which one it is.

121

The Highest Common Factor of two numbers is 4 and their Lowest Common Multiple is 24.

What is the product of the two numbers?

What might the two numbers be?

122

These are the shadows cast by a solid shape when a light is shone on it from the top, from the front, and from one side. What shapes could the solid be?

TOP

FRONT

SIDE

123

In this figure we started with two similar triangles, PQR and P'Q'R', and found a kind of 'average' by joining PP', QQ' and RR' and finding their mid-points, X,Y,Z.

Prove that XYZ is similar to PQR and P'Q'R'.

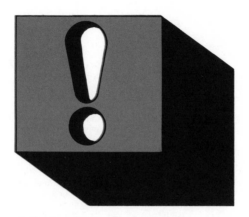

Hints for Solutions

Ideas and suggestions which can help, but which do not give away the solution.

HINTS FOR SOLUTIONS

1. Try cutting one triangle up to make the other triangle.
2. The solution is symmetrical, with just one horizontal line.
3. No Hint.
4. The best place to start is *b down*. This is a perfect square and a perfect cube, so it is a perfect sixth power. The only 2 digit sixth power is $2^6 = 64$.
5. Try old fashioned long-division.
6. Look at each line separately, especially the difference between each number and the next.
7. The areas of circles depend on the squares of their diameters.
8. The areas where they overlap take exactly the same area from the largest circle as they do from the smallest circles.
9. Trial and error. Use a calculator for the second and third — or quadratic equations.
10. No Hint.
11. Think about the wheels.
12. Trace the diagram, and fill it in very accurately with diamonds. Crude but it works very well.
13. No Hint.
14. Keep the hexagons as close together as possible.
15. One of the pentagons is a hole, an empty space.
16. No Hint.
17. Try tipping the can.
18. The rows have got to add up to $8 + 9 + 10 + 11 + 12 + 13 = 63$, but the numbers 0 to 9 only add up to 45. So the three corner numbers and the central number which are each counted three times (instead of once only), must make up the extra 18.

19. Two different prime numbers multiplied together (such as 2 x 3) always have four factors. (1, 2, 3, 6).
20. Numbers have many factors when they have at least two different prime numbers which are repeated as often as possible.
21. Imagine that the black circles are odd numbers, the empty circles are even numbers and you are adding them.
22. No Hint.
23. When you count the number of prisoners in all the rows, the prisoners in the four corner cells are each counted twice.
24. By comparing the three lines, P must be 2 more than Q, which must be 2 more than R.
25. J is a three-digit number, but J —1 is only a two-digit number.
26. T'J stands for I'm and TU'C stands for It's, so every T stands for i and every J for m, and U stands for t and C stands for s and so on.
27. Start with a triangle and look for a symmetrical solution.
28. In the three triangles given, there are 1, 8 and 27 triangles.
29. It is impossible to fill the end of a rectangle only 3 squares wide, so the rectangle must be at least 4 squares wide and the total number of squares must be divisible by 3.
30. $7 \times 9 = 63$
    $77 \times 99 = 7623$
    $777 \times 999 = 776223$
    and so on.
31. No Hint.
32. No Hint.
33. The smallest rectangle uses only four pieces, two of each shape.

HINTS FOR SOLUTIONS

34. $(5) \div (3 + 2) = 1$

35. $78^3 = 474552$ and $79^3 = 493039$ so one of the factors must be less than 78 and a prime number.

36. 2 people shake once, 3 people shake 3 times, 4 people shake 6 times . . .

37. The largest piece must be placed at a surprising angle.

38. Compare the given sum with this sum: $1 + \frac{1}{3} + \frac{1}{9} + \frac{1}{27} + \frac{1}{81} + \frac{1}{243} + \ldots$

39. The maximum area requires curved fencing.

40. No Hint.

41. Get a glove and try it!

42. The width of the passage makes absolutely no difference at all.

43. No Hint.

44. Factorise the left-hand side.

45. Draw a quadrilateral ABDE and its diagonals. Remember this is a difficult problem.

46. No Hint.

47. No Hint.

48. Discover what each line ought to add up to by adding up all the numbers and dividing by 4.

49. Don't use a calculator. Each of the given numbers is a perfect 10th power.

50. Wrapping in two different directions uses the same area of paper, though in two different shapes.

51. Imagine moving round the string and counting as you go: 1 left, 2 left, . . . 5 left and 1 up, 5 left and 2 up, and so on . . .

52. This is one of the most interesting problems in the book and can be solved in many ways. One idea: think of the same grid being drawn on each map, to the appropriate scale . . . Another idea: think of what would happen if the large map turned into the small map by rotating and getting

smaller . . . Another idea: imagine that there is a matching point, call it something, P say, and think of similar triangles.

53. An exact number of pounds is just a number of pence ending in 00.

54. This is an easy problem to start, but to get far you need patience, or a programmable calculator or a small computer, or a lot of insight. Some of the answers seem paradoxical.

55. Find the sum of all the rows and columns — most of these are already multiples of 5.

56. No Hint.

57. The answer is something to do with equilateral triangles.

58. Write the numbers as prime factors multiplied together.

59. Middle method $123 = 999 - 876$
Third method $8775 = 10,000 - 1225$

60. No Hint.

61. Translate CD and FE to make a triangle with AB, but without moving AB at all. Then look at the figure you have drawn.

62. Trial and error, a small computer or algebra.

63. The problem is equivalent to $a + 2\sqrt{ab} + b = c$.

64. The difference of two squares is always the sum of consecutive odd numbers: for example $12^2 - 7^2 = 23 + 21 + 19 + 17 + 15$.

65. Where can 279 be put?

66. Consult an athletics coach.

67. Try it and see.

68. There are an infinite number even if you only consider vertices on an infinite strip of dots, three dots deep.

```
.  .  .  .  .  .  .  .  .  .  .
.  .  .  .  .  .  .  .  .  .  .
.  .  .  .  .  .  .  .  .  .  .
```

HINTS FOR SOLUTIONS

69. The solutions to the second and third parts are connected.

70. No Hint.

71. Compare the sum of the series with the next number in the series.

72. $1 + \frac{1}{4} + \frac{1}{16} + \frac{1}{64} + \frac{1}{256} + \ldots?$

73. To make the area as large as possible, the corners should be on a well-known curve.

74. What difference do the wheels in the middle make?

75. The mode is the mark which is scored most often; the median is the middle mark when they are arranged in order of size.

76. Parallel lines.

77. No Hint.

78. For each sequence, think about the differences between one number and the next.

79. Experiment with accurate drawings.

80. Concentrate on the A, because it occurs twice.

81. The numbers round the inside pentagon, from 3 — 1 to 5 — 3 add up to 4, 5, 6, 7, 8.

82. Try it and see — then wonder how you could work it out without drawing.

83. A number which divides each of two numbers exactly, also divides their sum and other more useful combinations of them . . .

84. First sum: how can the answer be more than 1000 but less than 2000? Third sum: the first blob in the answer is *not* a zero.

85. Compare the tessellation in the problem with a honeycomb.

86. No Hint.

87. No Hint.

88. You need this identity:
$(a - b)(a + b) = a^2 - b^2$.

89. With ingenuity you can try it and see!

90. A calculator won't help unless it shows more than 8 figures.

91. All the numbers are whole numbers.

92. No Hint.

93. Take out 67 and try to divide the rest of the numbers so that one set is 67 more than the other in total.

94. All the pieces you need are triangles.

95. The pattern is triangular.

96. No Hint.

97. Adults do not run like this.

98. One way to see a solution is to imagine what happens when one of the lines rotates.

99. Try it and see.

100. No Hint.

101. Start by moving to (2,1), (4,2), (6,3) . . . and then turn right.

102. If the alphabet had only three letters, there would be nine combinations of two of them.

103. The solution is symmetrical.

104. You will need a little algebra, for example:
$(a + b)^2 = a^2 + b^2 + 2ab$.

105. No Hint.

106. Try it and see!

107. Try it and see, and try turning the cube round.

108. No Hint.

109. Think of the number in pairs.

110. Provided the angles are correct, the length of the sides makes no difference to the answer.

111. You can start by choosing any two adjacent sides and completing one parallelogram.

112. The amount of water flowing through each pipe is proportional to its cross-sectional area.

HINTS FOR SOLUTIONS

113. F is a perfect square, 2F is a two digit number, but 3F is three digits.

114. Two of the faces of the shape have four sides.

115. The number in the middle of the 3 by 3 square must be 5 in the first diagram and 13 in the centre of the second diagram.

116. One way to solve this problem is to think about the angles of the triangles and quadrilaterals.

117. Compare one number with the next but one number in the sequence you get.

118. F'P stands for I'm, so every F stands for the letter i and every P for an m.

119. Only one answer to this problem. Try it and see!

120. Twice the average of two numbers is equal to the sum of the two numbers.

121. The HCF is the largest number which divides them exactly. The LCM is the smallest number they both divide into exactly.

122. Imagine that the three shapes given are actually made of cardboard and used as three sides of the solid shape.

123. Two hints: this is a very difficult problem — and it is connected in an obscure way to problems 79 and 108.

If at first you don't succeed, try, try again.
(W.E. Hickson)

If at first you don't succeed, try, try again.
Then quit. No point in being a damn fool about it.
(W.C. Fields)

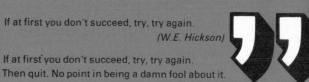

SOLUTIONS

The solutions are printed upside-down
to discourage all except the truly desperate.
See the quotations opposite.

117. The series gets closer and closer to
1—3—1—3—1—3 and so on.
The results for different starting
numbers or different operations we
leave to you. Of course, you would be
sensible to expect the same *kind* of
answer.

118. The longest word is TOGETHER.

119. For five faces meeting at each vertex,
think of a flat plane with lots of holes
in it. The rest is up to you.

120. R equals twice the sum of P and Q,
which is twice P and twice Q. But
twice P is Q + R. So R plus three
times Q still equals R! So Q must be
zero.

121. The product of the two numbers must
be 96. They *could* be 4 and 24 — but
there is another possibility also,
which we leave you to find out. Is
there a third possibility?

122. The hint should have solved this one.

123. A problem for investigation. No
solution given.

97. Small girl skipping along a pavement.
98. Many solutions, mostly depending on the fact that the angles of a quadrilateral add up to 360°
99. With the numbers given, the X number gets nearer and nearer to 8 and the Y number gets nearer and nearer to 4.
100. Just a further hint that there are two solutions to the first part and the answer to the last question depends on which you choose!
101. 14 moves.
    Problem
    How many other points may be reached in 14 moves but not fewer?
102. 67600 if 00 is counted as a two digit number; if 00 is excluded the answer is 66924.
103. No it is not. However, it is possible if the quadrilateral is a parallelogram.
    Problem: Why?
104. The sum of their squares is 84 and the sum of their reciprocals is $\frac{2}{5}$.
105. Three times as old.
106. Yes, it is. However, the minimum number of creases needed we leave to you. There is no need for the creases to overlap.
107. The slope must be very slightly under 45° if the cube is about to tip when placed with a face facing down the slope, as in our picture.
    However, if it is rotated so that an edge points down the slope, then the slope could be nearly 54° 44′.

108. The limit point is vertically above the right hand end of the first, horizontal, line.
    Problem
    Where would the limit point be if each line was still half the length of the previous line, but at an angle of 45° to it?
109. −50
110. Yes
111. Four ways, or eight ways if you count the reflections separately.
    Problem
    How many ways for a decagon with the same special property?
112. The single large pipe.
113. Reading across, top to bottom 8/27/88 : 891/49 : 3/441/8 : 66/726 : 44/21/3.
114. Make two squares, call them ABCD and PQRS. Join them with eight matches, like this:
    AP, AQ, BQ, BR, CR, CS, DS, DP.
115. Reading from top to bottom, left to right:
    19—1—2—20—23 :
    18—12—11—16—8 :
    21—17—13—9—5 :
    4—10—15—14—22 :
    3—25—24—6—7
    or any rotation or reflection.
116. No it is not.
    Problem
    How about if you were allowed to add just *one* new vertex on one side of the original triangle?

SOLUTIONS

80. First problem, reading across, top to bottom: 4/3, 2/9, 4,7.
Second problem, reading the same way: 4/9, 8/6, 7, 1.

81. The letters of MAGIC in that order should be:
12, 14, 11, 8, 10.

82. The nearest point you can reach is 32 along and 40 up, which is two to the right of the point you are aiming for.

83. 151

84. 225 × 7 = 1575 : 38 — 29 = 9 :
992 ÷ 8 = 124.

85. 6 : the reasons we leave to your investigation! What would the answer be if more than three polygons are allowed to meet at each vertex?

86. Your solution will check itself where the answers cross each other.

87. The problem does not make clear how the smallness of the differences is to be measured — so this is just one sensible arrangement: (reading anti-clockwise round the whole area starting at the far right-hand corner) 1, 2, 3, 5, 7, 10, 11, 9, 6, 4 and district 8 is the inside space.

88. Just a further hint: $9 - p^2$ must equal 4.

89. Neither: The oval is symmetrical — both ends are identical. It is called an ellipse. How can this be demonstrated?

90. $\sqrt{34782}$ is very slightly larger.

91. The numbers are 2, 3 and 5. Does it make any difference which is which?

92. Break the six bars in half and give a half to each person. Break the remaining four pieces in half again making eight quarters of a bar and give one piece to each person. Result six people get $\frac{1}{2}$ and $\frac{1}{4}$ each.
Problem
Can this method be used to share any number of bars between any number of people — or are there some numbers for which it would not work?

93. 67 + 31 + 4 + 5 = 107
46 + 24 + 19 + 17 = 106
Problem
How can you tell just by looking at the numbers that they cannot be divided into two parts whose sums are exactly equal?

94. Divide the fat parallelograms in the centre into halves: if you cut them the right way, five of these halves will make a pentagon. Cut the slim outside parallelograms in half the right way and the ten halves arranged round a point make a decagon.
Problem
Can the 12-sided and 14-sided figures and so on be dissected in the same way?

95. The rose bushes should form a tessellation of equilateral triangles. The largest number which can be fitted into the rose bed is 240.

96. When you are using an approximate method which gets as close to the answer as you need in a number of steps, then making a mistake may mean that you just need a few more steps — not that your answer is going to be wrong.

SOLUTIONS

63. There are many solutions, such as 8, 18 and 50: The square roots of these numbers are $2\sqrt{2}$, $3\sqrt{2}$, and $5\sqrt{2}$ respectively.
Problem
Are there any solutions in which the numbers a and b are not perfect squares *and* do not have any common factor?

64. All numbers can be written as a difference of two squares except multiples of 4.

65. Reading across from top to bottom 2/7135 : 81/279 : 1/431/6 : 517/71 : 4672/2.

66. For discussion.

67. See the hint.

68. Here are the co-ordinates for the first few solutions, using just a strip of the grid as in our hint.

(0,0) (0,0) (0,0) (0,0)
(1,0) (1,1) (1,2) (1,3)
(2,3) (2,5) (2,7) (2,9)
and so on . . .
There are an infinity of other solutions.

69. First part; draw a diameter and join the two halves: second part; cut in straight lines from one corner to the next corner but one in each direction: third part; cut off three pieces as in the second part, and fit the three pieces together to make the second equilateral triangle.

70. I am nine years old.

71. Work out what would be the next term in the series. Take away one. Halve the answer.

72. Two-thirds.

73. The area is a maximum when the four corners lie on a circle, and the area is 16 units.

74. 16 times a minute. If the smallest wheel rotates once per second or 60 times per minute, its diameter must be 8 cms.

75. 5, 5, 4, 1 and 0.

76. Draw a line parallel to AB, so that it meets the same two lines through X. Through the points where it meets these two lines, draw lines parallel to AD and BC. Join the points where these meet XD and XC and you have another quadrilateral similar to ABCD.
Problem
Can this method be used to blow up or scale down any straight-line drawing?

77. Reading across, top to bottom 969/6 : 48/96 : 58/65 : 6/756.

78. First sequence: 127, 255, 511.
Second sequence: 56, 84, 120.
Third sequence: 85, 171, 341.
Fourth sequence: 31, 57, 105.

79. An infinite number! If you draw several of them accurately you will find a lot of possibilities to investigate — and a connection with problem 123. We leave those problems up to you!

SOLUTIONS

42. 2⅔ metres.
Problem
How far from each vertical wall do the ladders cross?

43. Just 4 children.

44. $s(B — s) = (B — s)(B — (B — s))$.
Therefore, if $s(B — s) = A$ so does $(B — s)(B — (B — s)) = A$ and so $B — s$ is another solution.

45. ABDE is the quadrilateral drawn initially: The parallelogram CXFY has been drawn round it, so that its sides are parallel to the diagonals AD and BE. The six points ABCDEF are one solution, AXBDYE are another. However to find *all* possible solutions is more tricky: the points must be chosen so that
$$\overline{AB} + \overline{CD} + \overline{EF} = \overline{BC} + \overline{DE} + \overline{FA}$$

46. "It is not true that every quadrilateral is either a parallelogram, a rectangle or a kite" is one possible answer.

47. 5 moves.

48. Exchange 5 and 6, 7 and 8, 1 and 4.

49. 2^{30} is less than 3^{20}.

50. Rectangles 80 x 55 or 110 x 40 will each be sufficient and the area of each rectangle is 4400. (You will need tape to seal the paper at the ends).
Problem
Is it possible to use less paper by cutting it, or wrapping the paper at an unusual angle?

51. 3 to complete the loop; 5 to get back to the beginning.

52. See the hints. You can check your answer by measurement.

53. $1/100$ if $1/2$p are not allowed, otherwise $1/200$. (So the clerk can expect this to happen either 2½ or 1¼ times a year on average, if he works 250 days a year.)

54. See the hint.

55. Cross out the first 3 in the second row and the 6 in the third row.

56. David Smith, Peter Jones and John Brown.

57. The sides of the rectangle are in the ratio $\sqrt{3} : 1$ or approximately $1 \cdot 73$ to 1.

58. Because the factors of perfect squares always come in pairs, but the numbers in the sequence always have a single factor 3.

59. A partial solution only: the middle method is a way of counting up from 876 to 1225 stopping at the convenient number 999 on the way. One way to think about the factors of the third method is
$A — B = 10,000 — ((10,000 — A) + B)$

60. The solutions are self-checking.

61. The hint should be sufficient.

62. Five solutions for the first problem if 0 counts as a square:
10, 11, 13, 18, 35.
The second problem has only two solutions 6 and 22.

SOLUTIONS

28.

This is the next figure and it contains
64 triangles, so the pattern looks like
this:

1   2   3   4   ?   ?   ?
1³ 2³ 3³ 4³ ?   ?   ?

Can you prove that this pattern always
gives the correct answer?

29. The smallest rectangle is 4 by 6.

30. 777776222223.
Problem
Can you predict 222222 x 999999 in a
similar way?

31. This number has many special features
and we will mention only the simplest.
If you multiply it by any number from 1
to 16 you get the same sequence of
digits, but starting at a different point!
If you multiply it by 17 you get
5999999999999994; if you add the
second half of the sequence to the first
half, you get 99999999.
Problem
Investigate the sequence 142857
which has many similar properties,
although it is shorter and simpler!

32. Your answer will check itself — by
measurement.

33. Place a U shaped piece at either end of
a 3 x 6 rectangle and fill the middle
with two Z pieces. The second question
should not now be too difficult. You are
on your own for the third part which is
very hard.

34. There are several different solutions.
Our hint was a clue to this one:
$(9 + 8 — 7 — 6 + 5 — 4) \div (3 + 2) = 1$.

35. 479491 = 53 x 83 x 109.

36. 10 people.

37. Place the longest sides of the three
small triangles against one of the
longest sides of the largest piece and
half of the other longest side — then
add the fifth piece in an obvious way.

38. Mathematicians say that the sum is ½.
They *mean* that the sum gets as near to
½ as you like if you go on long enough,
and never goes beyond a ½. For
example if you continue the sum far
enough it will add up to more than
0·4999999999 but it will never add up
to as much as 0·50000000001.

39. The maximum area requires a curved
fence in the shape of half a circle.
Problem
If the fence has to be used instead to
make three sides of a rectangle with as
large an area as possible, what shape
should the rectangle be?

40. 39; it is just possible that the number 1
record could drop out of the charts
completely and all the other records
move up exactly one place each —
possible, but unlikely.

41. Assuming that the hand and fingers of
the glove are naturally a bit curved to fit
the hand inside, then a left-hand glove
becomes a right-hand glove.
Problem
If you look at a right-hand glove in a
mirror, will it look like a right-hand
glove or a left-hand glove?

15. The hint should be sufficient.

16. There are about 40 generations in 1000 years and the number of ancestors apparently doubles with each generation back, so the total should be 1,099,512,000,000 to the nearest million. The total population then was only about 340,000,000 and even today is only about 4,500,000,000.

17. The solution is obvious, once you can see it!

18. Once you have discovered which numbers fill the corner circles and the centre circle there are many ways to fit them into the diagram. Here is one way: Going round the outside triangle, from the top circle:
3—1—4—9—0—8—3; joining the corner circles to the centre:
3—5—2, 4—6—2 and 0—7—2.

19. The numbers are the squares of prime numbers.
$2^2 = 4$, $3^2 = 9$, $5^2 = 25$ and $7^2 = 49$.
Problem
Can you find the only two numbers less than 100 which have five factors and the one number less than 100 which has seven factors?

20. 72 and 96 each have 12 different factors including themselves and 1.
Problem
Can you find the one number less than 1000 which has no less than 27 factors?

21. This is a very rich investigation with many possibilities.
The rule is that two circles of the same colour have a white circle below and between them; two different circles have a black one.
The pattern will repeat eventually, but we leave you to investigate how!
It is possible to predict the bottom circle from the top row if the top row has 3, 5, 9 . . . circles in it. Otherwise prediction is possible but very complicated — more to investigate.

22. Even if you talk very quickly, we estimate it would take about 30 days to count from 1 to 1,000,000 — and much longer if you get tired and start making mistakes.

23. With the hint you should be able to find one of several possible solutions.

24. P = 7, Q = 5, R = 3.

25. *Across* A: 99, C: 24, F: 144, H: 77, J: 100, M: 13, N: 540, Q: 52, R: 25.
Down b: 94, d: 470, e: 27, f: 12, g: 405, j: 135, k: 10, m: 15, p: 42.

26. The longest word is BUTTERFLY — and the rest is up to you.

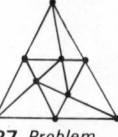

27. *Problem*
Now can you arrange 10 points so that each point lies on 3 straight lines and there are three points on every line. Rather harder!

1. Follow the hint and you will find they have the same area. There are of course other methods.
Problem
Can you find another pair of triangles with equal areas, the same slanting sides but different bases?

2. No solution given, but we assure you that there is a perfectly fair answer. It is just a question of finding it.

3. The first square can be completed easily and the second not at all. The pair of numbers between 4/1 and 4/2 can never make them both add up to 10. The third part can be done in many different ways and makes an interesting investigation. The last problem is also for you to decide!

4. *Across* A: 16, C: 196, F: 144, G: 64, H: 38416, L: 32, M: 172, P: 280, Q: 84.
Down a: 113, b: 64, c: 14410, d: 961, e: 64, j: 828, k: 624, l: 32, n: 78.

5. 7 goes into 999,999 exactly 142857 times, and then into 999,999,999,999 and into 999,999,999,999,999,999 and so on!
13 divides exactly into 999,999 and so on and 17 into 9,999,999,999,999,999 and so on. 24 *never* divides into a number composed of 9's.
444,444 and so on.

6. Reading downwards the next three numbers are —37, 43 and 11.
Problem
What happens if the triple sequence is carried backwards, from 3—8—6 to the left?

7. The areas of the outer circle, the inner circle and the space between them are, respectively
$\frac{49\pi}{4}$, $\frac{25\pi}{4}$ and $\frac{24\pi}{4}$
so the inner circle is larger.
Problem
Can you find another pair of circles, whose diameters are whole numbers, so that the difference between the inner circle and the space between is $\frac{\pi}{4}$?

8. They are equal.

9. 3 and 9, 2·3 and 9·7. The last part can only be solved exactly by using square roots. They are
$6 - \frac{1}{2}\sqrt{56}$ and $6 + \frac{1}{2}\sqrt{56}$
or approximately
2·258343 and 9·741657

10. THINKER can be spelt 20 ways.
Problem
In how many ways could it be spelt if you do *not* have to go to the next letter along a line — for example you could go from either of the H's to any of the I's.

11. See the hint.

12. A regular 14 sided figure divides up into 21 diamonds, and a regular 6 sided figure into 3 diamonds. The rest we leave you to investigate!

13. Every answer cross-checks.

14. These six pieces make two hexagons and the space in the centre can be filled with whole hexagons.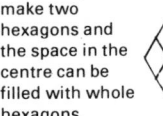

SOLUTIONS
AND SOME
FURTHER PROBLEMS
& INVESTIGATIONS